LoRa
MESH NETWORKING
FOR OFF-GRID
COMMUNICATION

A FIELD MANUAL

FOR MODERN MESH NETWORKING, EMERGENCY PREP, AND DE-CENTRALIZED COMMUNICATION

Tony Vortex

First Edition
ISBN 978-1-949432-32-9

Published by:

Inner Alchemy's Publishing
332 S. Michigan Ave.
Ste 121-C141
Chicago, IL 60604-4434

Printed in the United States of America

Table of Contents

Introduction 4

Deployment Reference Guide 9

LoRa in Mesh Networks.......................................13

Antenna's...15

What is a LoRa Gateway? 21

Ways to Use Both Antennas on a Single LoRa Device...................................24

Device Roles 27

Stationary Use Cases 34

Alternative Use Cases..36

RAK Device VS LoRa Meshtastic Device 39

LoRaWAN vs Meshtastic..42

Encryption on Meshtastic Networks 45

Recommended tools and equipment
for various grid down scenarios........................46

Tactical Operations Center.................................49

Glossary 53

Introduction

As someone who's been deeply entrenched in LoRaWAN, Meshtastic, amateur (ham) radio, and wireless communications for over a decade, I've seen firsthand how rapidly these technologies are evolving—and how invaluable they've become in real-world deployments, especially where traditional networks fall short.

This guide is not just a spec sheet—it's a **strategic blueprint** for understanding, deploying, and optimizing **off-grid mesh networks** using **LoRa-based devices**. Whether you're building infrastructure for emergency comms, rural IoT sensor grids, or hobbyist mesh chat networks, this document is your field manual.

✪ What Are These Devices?

At their core, **LoRa devices** use a long-range, low-power wireless protocol called **LoRa** (Long Range), which operates in the unlicensed ISM bands (e.g., 915 MHz in North America). Unlike Wi-Fi or cellular, LoRa is designed for sending small amounts of data over **very long distances** using **very little power**.

Meshtastic is an open-source firmware that turns LoRa-capable devices (like the TTGO T-Beam or RAK4631) into mesh network nodes capable of **peer-to-peer, decentralized communication**. These nodes don't require internet, cellular, or infrastructure—just power and proper configuration.

⚡ How It All Connects

There are three key players in this ecosystem:

- **LoRa** – The physical modulation technology.

- **LoRaWAN** – A structured protocol for connecting sensors to internet gateways.

- **Meshtastic** – A mesh protocol that allows devices to chat and relay messages without infrastructure.

You flash Meshtastic onto a LoRa-capable device, and suddenly that device becomes part of a secure, decentralized, off-grid communication mesh.

🌐 Where Are These Used? And Why?

Let's break down **use-case scenarios**, and why you'd choose **Meshtastic/LoRa** over Wi-Fi, Cellular, or Ham Radio:

1. Emergency Preparedness & Disaster Response

When the grid goes down, your network shouldn't.

- Use rugged nodes on rooftops to create a communication backbone.
- Portable T-Beams allow teams to stay in touch via encrypted chat.
- Deploy SOS beacons and GPS trackers for rescue or medical teams.

🔧 Why LoRa/Meshtastic?

No internet or cell towers needed. Long battery life. Works off-grid for weeks or months.

2. Off-Grid Communities or Rural Farms

You don't need cell towers to stay connected.

- Track equipment, vehicles, or livestock with GPS-enabled nodes.
- Set up weather sensors and environmental monitors using RAK boards.
- Use mesh chat to coordinate across wide rural areas without cellular service.

🔧 Why LoRa/Meshtastic?

Huge range (5–10 miles). Solar/battery powered. Minimal data needs.

3. Prepping, Hiking, or Expedition Teams

Keep your team connected deep in the backcountry.

- Each member carries a T-Beam unit.

- Send location pins, messages, or emergency broadcasts.

- Base camp can act as a gateway or relay point.

🔧 Why LoRa/Meshtastic?

No monthly service. Encrypted. Lightweight gear.

4. Event Mesh Networks

Festivals, outdoor expos, or survivalist meets.

- Create local-only chat rooms.

- Broadcast announcements.

- Monitor safety zones or access points with geofencing.

🔧 Why LoRa/Meshtastic?

Build a private digital bulletin board. No IT infrastructure needed.

5. IoT Sensor Grids

Smart farming, weather stations, or industrial telemetry.

- RAK4631 + WisBlock lets you build low-power sensor nodes.

- Integrate with Meshtastic or LoRaWAN depending on application.

- Use the Pi bridge for pushing data to dashboards.

🔧 Why LoRa/Meshtastic?

Insanely efficient. Battery/solar ready. Minimal configuration overhead.

📡 Antennas, Gateways & Range

This guide covers:

- Antenna types (Omni vs Yagi vs Parabolic)

- Mounting strategies

- When to use masts, grounding, or telescoping poles

- Line-of-sight calculators and tools for reliable long-distance comms

You don't just "plug in an antenna"—you design your signal path.

Choosing the right antenna and elevation is what transforms a good mesh into a **great one**.

🔐 Encryption: Why It Matters

Security isn't optional—it's foundational.

Meshtastic supports **AES-256 encryption** between nodes. This ensures that even if someone is listening on the same frequency, they cannot decode your messages. Each node can use a shared channel key or individual keys for group communication.

This makes Meshtastic ideal for:

- Covert installations

- Emergency networks

- Sensitive team coordination

⚒️ Device Roles & Network Planning

The guide teaches how to configure each node's role:

- **Client** – Regular user device.

- **Router** – High elevation relay.

- **Repeater** – Pure message forwarder.

- **Sensor/Tracker** – For telemetry and movement tracking.

- **Gateway** – Optional internet bridge (via Wi-Fi or Pi).

Using the right device in the right role avoids redundancy, saves power, and optimizes performance across the network.

🧠 Final Thought from the Field

If you're building a mesh for fun, survival, research, or community empowerment—**Meshtastic gives you professional-grade capabilities without needing a tower lease or a data plan**.

But it's not plug-and-play. It's plug, configure, mount, align, tune, and optimize. That's what this guide helps you do—with clarity and precision.

Deployment Reference Guide

1. Overview

This technical reference guide provides quick-access information for deploying and optimizing LoRa and Meshtastic-based mesh networks, including antenna selection, hardware recommendations, deployment strategies, and practical field setups. Ideal for off-grid communication, IoT integration, and emergency preparedness.

2. LoRaWAN vs Meshtastic

Quick Comparison:

Feature	LoRaWAN	Meshtastic
Primary Use	IoT sensor data collection	Off-grid mesh messaging
Topology	Star (node -> gateway)	Mesh (node <-> node)
Gateway Required	Yes	No
Internet Required	Usually	Optional
Peer Messaging	Limited	Full messaging/chat support
Encryption	AES-128	AES-256 (optional)

3. Recommended Hardware

• TTGO T-Beam (ESP32 + GPS): Ideal for mobile messaging and tracking.

• RAK4631 + WisBlock Base: Best for sensor nodes, low-power routers, or solar-powered setups.

• Raspberry Pi: Acts as a serial gateway for internet bridging via USB connection.

4. Antenna Selection

• Blade Antenna: Default included antenna.
Good for testing, short range use.

• 915MHz 8dBi Omni Antenna (Fiberglass):
Great all-around long-range antenna with 360° coverage.

• 10–13dBi Yagi Antenna: Directional.
Ideal for point-to-point, long-distance links.

5. Antenna Mast Recommendations

• Fiberglass Telescoping Mast
  – Non-conductive, lightweight, RF-transparent.

• Chain-link Fence Top Rail – Cheap, strong, locally available.

• TV Antenna Mast – Designed for vertical installations. Works well with roof brackets.

• Accessories: U-bolts, Guy Wires, Grounding Clamp + Rod.

6. Strategic Mesh Layout

Use low-power, stable nodes (RAK4631) on rooftops or high elevation points for backbones. Deploy ESP32-based T-Beams for mobile users. Combine with a Raspberry Pi gateway bridge if internet connectivity is required.

7. Internet Gateway and Mesh Map Reporting

• RAK4631 + RAK1920 WiFi Module can connect directly to internet.

• Alternatively, use Raspberry Pi + USB Meshtastic device as a bridge.

• Enable GPS, telemetry, and position_broadcast_secs in firmware to appear on Meshtastic Map.

8. ESP32 vs nRF52840 (RAK4631)

Category	ESP32 (T-Beam)	nRF52840 (RAK4631)	Best For
Power Efficiency	Moderate	Excellent	Long-term sensor nodes
WiFi Support	Yes	No	Gateways
Bluetooth	Yes (higher power)	Yes (low power)	BLE-only nodes
GPS	Built-in on T-Beam	Requires add-on	Tracking
Cost	Lower	Higher	Budget vs Quality

9. Deployment Role Examples

• Repeater on Hilltop – RAK4631 with solar + omni antenna.
• Tracker – T-Beam sending GPS every 120s, role: tracker.

• Sensor Station – RAK4631 + I2C sensors, role: sensor.

• Emergency Beacon – T-Beam in repeater mode with pre-set SOS broadcast.

• Gateway Bridge – ESP32 or RAK4631 + Raspberry Pi, connected to WiFi/Ethernet.

10. Tools and Configuration

• Flashing Firmware: https://flasher.meshtastic.org

• Meshtastic CLI: https://github.com/meshtastic/Meshtastic-python

• Map Viewer: https://map.meshtastic.org

• Line-of-Sight Tool: https://www.scadacore.com/tools/rf-path/rf-line-of-sight/

• Serial Setup: `meshtastic --set <parameter> <value>`

LoRa stands for **"Long Range"** and is a low-power, long-range wireless communication protocol developed by Semtech. It is designed for **low-bandwidth, long-distance communication** between IoT (Internet of Things) devices, sensors, and gateways.

LoRa in Mesh Networks

LoRa itself typically operates in a **star topology** (point-to-multipoint) rather than a mesh. However, **LoRa Mesh Networks** can be implemented using protocols that extend LoRa's capabilities, such as:

1. **LoRa Mesh (LoRaMesher, Meshtastic, The Things Network)**

   o   Some systems integrate **mesh routing** to allow nodes to relay messages beyond their direct communication range.

   o   **Meshtastic** is an example of an open-source project using LoRa for **off-grid, decentralized mesh networking.**

2. **LoRaWAN (LoRa Wide Area Network)**

   o   LoRaWAN is a **network protocol** that builds upon LoRa to enable **low-power, long-range communication** between devices and gateways.

   o   Unlike true mesh networks, LoRaWAN uses **star-of-stars topology**, where gateways relay data to a central network server.

Key Features of LoRa Mesh Networks

- **Ultra-low power consumption** – great for battery-operated devices.

- **Long-range communication** – up to **10+ miles in rural areas** and **2-5 miles in urban environments.**

- **Decentralized and off-grid** – useful in areas with no internet or cellular coverage.

- **Low bandwidth** – best for **text messages, sensor data, or telemetry**, not for high-speed applications.

Meshtastic, which uses **LoRa (Long Range) technology**, operates on different **unlicensed ISM (Industrial, Scientific, and Medical) radio**

bands worldwide. The frequencies vary by region due to **government regulations**. Below are the main **LoRa frequencies used for Meshtastic worldwide**:

Meshtastic LoRa Frequency Bands by Region

Region/Country	Frequency Band	Notes
North America (USA, Canada)	915 MHz (902-928 MHz)	ISM band, FCC-regulated. Commonly used for LoRaWAN and Meshtastic.
Europe (EU, UK)	868 MHz (863-870 MHz)	ETSI-regulated. Power and duty cycle restrictions apply.
Australia & New Zealand	915 MHz (915-928 MHz)	Similar to North America but with local regulations.
India	865 MHz (865-867 MHz)	Lower power limits than EU.
China	470 MHz (470-510 MHz)	Different from most other regions, unique to China.
Japan	920 MHz (920-925 MHz)	Strict regulation on power and duty cycle.
South Korea	920 MHz (920-923 MHz)	Similar to Japan, regulated for specific applications.
Russia	868 MHz (864-865 MHz)	Similar to EU but stricter government oversight.
South America (Brazil, Argentina, etc.)	915 MHz (902-928 MHz)	Matches North America, ANATEL-regulated in Brazil.
Africa & Middle East	868 MHz or 915 MHz	Depending on the country's telecom regulations.

Important Notes:

- **Duty Cycle Restrictions:** In Europe and other areas, LoRa transmissions must adhere to **duty cycle limits**, meaning nodes can only transmit for a fraction of the time.

- **Power Limits:** Some regions, like **Japan and the EU**, restrict transmission power more than others.

- **Meshtastic Customization:** Meshtastic firmware allows setting the correct **regional frequency** to comply with local laws.

What Kind of Pole Should Be Used for Mounting a LoRa Antenna?

To maximize **LoRa signal propagation**, the antenna should be placed **as high as possible** with minimal obstructions. Here's what to consider when selecting a pole:

Best Pole Options:

- **Fiberglass Mast:** Lightweight, corrosion-resistant, and doesn't interfere with radio signals.

- **Telescoping Aluminum Pole:** Adjustable height and strong, but may require grounding.

- **PVC Pipe (Temporary Use):** Non-conductive and inexpensive, but less sturdy for permanent installations.

- **Galvanized Steel Pipe:** Very durable, but heavier and may require grounding.

- **Telescoping Flagpole:** A good DIY option for extending antennas to high elevations.

Additional Considerations:

- **Mounting Brackets:** Use **U-bolts or pole clamps** to secure the antenna.

- **Guy Wires:** If going over **20 feet**, consider using **guy wires** to stabilize the pole.

- **Non-Conductive Material:** If near a **metal roof or other structures**, fiberglass or PVC helps prevent interference.

- **Grounding (for Metal Poles):** If using a **metal pole**, install a **lightning arrestor** and ground the pole to prevent damage.

To establish a reliable 6-mile LoRa link between two locations, selecting the appropriate antennas and ensuring a clear line of sight (LOS) are crucial. Below are specific antenna recommendations and tools to verify LOS:

Antenna Recommendations

1. **Embedded Works 915MHz 13 dBi Yagi LoRa Antenna**

   o **Description:** This high-gain Yagi antenna is optimized for 915MHz LoRa applications, offering a 13 dBi gain to enhance signal strength and range.

   o **Features:**

      - Frequency: 915MHz

      - Gain: 13 dBi

      - Connector: N-type Female

      - Length: 820mm

   o **Link:** Embedded Works 915MHz 13 dBi Yagi Antenna

2. **SIGNALPLUS Outdoor Yagi 915MHz LoRa Antenna**

   o **Description:** Designed for outdoor use, this 10 dBi Yagi antenna is suitable for LoRa miners and long-distance communication.

   o **Features:**

      - Frequency: 824-960MHz (covers 915MHz)

      - Gain: 10 dBi

      - Includes 30ft RG58 RP-SMA cable and SMA adapter

      - Length: 0.7 meters

   o **Link:** SIGNALPLUS Outdoor Yagi 915MHz Antenna

3. **Outdoor 915MHz Parabolic Grid Directional Antenna**

    o **Description:** This high-gain parabolic grid antenna is suitable for long-distance LoRa communication, offering a 16 dBi gain.

    o **Features:**

        ▪ Frequency: 915MHz

        ▪ Gain: 16 dBi

        ▪ Designed for outdoor use with robust construction

    o **Link:** Outdoor 915MHz Parabolic Grid Antenna

Line-of-Sight Verification Tools

1. **SCADACore's RF Line-of-Sight Tool**

    o **Description:** This free online tool calculates the line-of-sight between two points by considering antenna heights and Earth's topography. Users can input coordinates or drag markers on a map to visualize elevation profiles and potential obstructions.SCADACore+1SCADACore+1

    o **Link:** SCADACore RF Line-of-Sight Tool

2. **everythingRF's Line of Sight Calculator**

    o **Description:** This calculator determines the theoretical line-of-sight distance based on antenna heights, considering Earth's curvature. It's useful for estimating the maximum range between two antennas.WinRFCalc RF Calculator

    o **Link:** everythingRF Line of Sight Calculator

3. **RF elements' Link Calculator 2.0**

    o **Description:** This tool estimates link performance by considering various parameters, including antenna types, frequencies, and distances. It helps in planning and optimizing point-to-point links.

- o   **Link:** RF elements Link Calculator 2.0

Additional Considerations

- **Antenna Alignment:** Ensure that directional antennas (like Yagi or parabolic) are precisely aimed at each other to maximize signal strength.

- **Antenna Height:** Mount antennas at sufficient heights to clear obstacles and maximize the Fresnel zone clearance.Wikipedia

- **Cabling:** Use low-loss coaxial cables and keep them as short as possible to minimize signal attenuation.

If you want your **LoRa device to broadcast in every direction** as far as possible within its location, you'll need an **Omni-Directional Antenna** instead of a Yagi or parabolic antenna. These antennas radiate signals **360° horizontally**, making them ideal for **mesh networks**, **LoRa gateways**, and **area-wide coverage**.

Best Omni-Directional Antennas for Maximum Coverage

1. RAKwireless 5.8dBi LoRa Omni Antenna (915MHz)

- **Best for:** General long-range broadcasting in all directions.

- **Gain:** 5.8 dBi (good balance between range and coverage).

- **Connector:** N-Type Female.

- **Mounting:** Includes pole-mounting bracket.

- **Pros:** Good for **mesh networks, relay nodes, or fixed LoRa gateways**.

- **Cons:** Mid-range gain; higher gain antennas may be needed for extreme range.

- **Link:** RAK 5.8dBi LoRa Antenna

2. Laird Technologies 915MHz 8dBi Omni Antenna

- **Best for:** Maximum range in urban and rural environments.

- **Gain:** 8 dBi (higher gain = longer range but slightly narrower vertical beam).

- **Connector:** N-Type Female.

- **Mounting:** Requires a pole or mast.

- **Pros: Excellent for LoRaWAN gateways** in a central location.

- **Cons:** Higher gain means **less coverage directly below the antenna** (mount it high!).

- **Link:** Laird 8dBi LoRa Omni Antenna

3. McGill Microwave 915MHz 10dBi Omni Antenna

- **Best for: Absolute maximum broadcast range**.

- **Gain:** 10 dBi (highest practical omni gain before needing a directional antenna).

- **Connector:** N-Type Female.

- **Mounting:** Needs a sturdy mast due to wind loading.

- **Pros:** Ideal for **covering a large area from a single point**.

- **Cons: Very narrow vertical beam** – only good for broadcasting outward at a distance.

- **Link:** McGill Microwave 10dBi LoRa Antenna

Optimizing for Maximum Range

To get the most out of an **omni-directional LoRa antenna**, follow these guidelines:

1. Mount it as High as Possible

- Elevate the antenna **30-50 feet above obstacles** (buildings, trees, etc.).

- Use a **fiberglass or aluminum mast** for the best height without interference.

2. Use Low-Loss Coaxial Cable

- If your LoRa device is far from the antenna, use **LMR-400 or LMR-600** to minimize signal loss.

- Keep the cable **as short as possible**.

3. Choose the Right Gain

- **Lower Gain (3-5dBi): Best for hilly terrain** or if you need coverage **above and below** the antenna.

- **Mid Gain (5.8-8dBi):** Balanced for **flat and urban areas**.

- **High Gain (10dBi+):** Best for **maximum range on flat open land**, but weaker close-range coverage.

4. Check Regional Frequencies

- **North America:** 915MHz

- **Europe:** 868MHz

- **China:** 470MHz

- Ensure the antenna you purchase matches your LoRa frequency band.

Final Recommendation

If you want **maximum area coverage** with **no specific point-to-point link**, go with an **8dBi omni-directional antenna** like the **Laird 8dBi** or **McGill 10dBi**.

What is a LoRa Gateway?

A **LoRa Gateway** is a **central communication hub** that **receives and transmits** LoRa signals between **LoRa end devices (nodes)** and a **network server**. It acts as a bridge, forwarding messages from multiple LoRa devices to the internet (if connected) or to other devices in a **private LoRa network**.

How Does a LoRa Gateway Work?

1. **LoRa Nodes (Sensors or Communicators)** transmit data using **LoRa radio signals**.

2. **The LoRa Gateway** receives signals from multiple nodes within its range.

3. The gateway **forwards the data** to a **network server** via:

    o **Ethernet (Wired)**

    o **Wi-Fi**

    o **Cellular (4G/LTE)**

    o **Satellite (for remote locations)**

4. The network server processes the data and **sends it to an application or dashboard**.

Types of LoRa Gateways

There are two main types of LoRa gateways:

1. Packet Forwarder Gateway (Basic LoRa Gateway)

- Acts as a **relay station** for LoRa signals.

- Does **not process data**; simply **forwards packets** to a cloud-based or private network server.

- Example: **RAK2245 Pi HAT LoRa Gateway**

2. LoRaWAN Gateway (Smart Gateway)

- Fully supports the **LoRaWAN protocol**,
  allowing multiple devices to communicate efficiently.

- Can work with **public networks** like
  The Things Network (TTN) or **Helium.**

- Example: **RAK7258 LoRaWAN Gateway**

Why Use a LoRa Gateway?

✔ **Long-Range Communication**
  – Can cover **miles** with low power.

✔ **Connects Multiple Devices**
  – One gateway can handle **thousands of nodes**.

✔ **Internet or Local Network Access**
  – Can be used with or without internet.

✔ **Scalable**
  – Ideal for **smart cities, agriculture, and emergency networks**.

Recommended LoRa Gateways

For Home or Small Networks:

1. **RAK7258 Indoor LoRaWAN Gateway**

    ○ Frequency: **868/915MHz**

    ○ Interfaces: **Wi-Fi, Ethernet, Cellular (optional)**

    ○ Link: RAK7258 Gateway

For Long-Range & Outdoor Use:

2. **RAK7249 Industrial Outdoor LoRaWAN Gateway**

   o   Weatherproof, long-range **multi-channel gateway**.

   o   Uses **PoE (Power over Ethernet)** for easy installation.

   o   Link: RAK7249 Gateway

3. **Helium Hotspot LoRa Gateway**

   o   Can be used for **private LoRa networks** or **Helium mining**.

   o   Link: Helium LoRa Gateway

Do You Need a LoRa Gateway?

- **Yes, if** you want **nodes to communicate over a large area** and need **internet connectivity**.

- **No, if** you are using **only peer-to-peer (Meshtastic) LoRa nodes**, where no centralized gateway is needed.

Can an omni-directional and a directional antenna be used on the same Lora device?

ChatGPT said:

Yes, an **omni-directional and a directional antenna** can be used on the **same LoRa device**, but it depends on the **hardware capabilities** and the **purpose** of the setup. Here's how it can be done and the considerations to keep in mind.

Ways to Use Both Antennas on a Single LoRa Device

1. Dual-Antenna Setup with a LoRa Device That Supports Multiple Antennas

- Some **advanced LoRa gateways and transceivers** have **multiple antenna ports** (e.g., **diversity antennas or separate TX/RX ports**).

- Example: The **RAK7249 Outdoor LoRaWAN Gateway** supports multiple antennas.

- **Use Case:**

  - The **omni-directional antenna** covers a **local mesh** or general area.

  - The **directional antenna** is used for **point-to-point long-range communication**.

2. Using an Antenna Switch (Manually or Automatically)

- If the **LoRa device has only one antenna port**, you can use an **RF antenna switch** to select between an **omni-directional** and a **directional antenna**.

- **Use Case:**

  - If the node is stationary, you can manually switch between antennas for different tasks.

  - Some advanced systems can automatically switch based on signal conditions.

3. Using a Duplexer or RF Combiner

- A **duplexer** or **RF combiner** allows two antennas to connect to one port, but there is **signal loss**.

- **Use Case:**

    - Mostly used in radio applications, **not ideal for LoRa**, since it can cause **interference** and **weaken signals**.

When Does This Setup Make Sense?

✔ **Yes, Use Both Antennas If:**

- You want **local coverage** (omni) while maintaining a **long-range link** (directional).

- Your LoRa device has **dual-antenna ports** or you use an **antenna switch**.

- You're running a **hybrid mesh + point-to-point network**.

✘ **No, Avoid Using Both If:**

- Your LoRa device **only has one antenna port** (without an external switch).

- The setup **causes interference** or **reduces efficiency** (e.g., signal loss in a combiner).

Recommended Devices for Dual Antenna Setup

1. **RAK7249 Outdoor LoRaWAN Gateway**
   – Has multiple antenna ports.

    - RAK7249 Link

2. **Antenna Switch Example:**

    - Coaxial RF Antenna Switch

Final Recommendation

- If you **only have one antenna port**, choose an **omni for general coverage** OR a **directional for long-range**.

- If your **LoRa device supports multiple antennas**, you can **optimize coverage and range** by using both.

Device Roles

In Meshtastic, devices can be assigned various roles to define their behavior within the mesh network. Below is a table summarizing each role, its use case, and the associated pros and cons:

Device Role	Description	Use Case	Pros	Cons
Client	Default role for general-purpose devices. Capable of sending, receiving, and routing messages.	Suitable for most users, especially in personal or non-critical applications.	- Flexible and versatile. - Participates in message routing.	- May consume more power due to active participation in routing.
Client Mute	Similar to Client but does not route messages from other devices.	Ideal for mobile or portable devices where conserving battery and reducing network traffic is important.	- Conserves battery life. - Reduces network congestion.	- Does not assist in extending network coverage.
Router	Primarily routes messages to extend network coverage. Typically stationary and placed in strategic locations.	Best for fixed installations in high, unobstructed locations to act as network hubs.	- Enhances network reliability and range. - Prioritizes routing over own messages.	- Requires optimal placement; improper use can cause network issues.
Repeater	Similar to Router but does not broadcast its own telemetry; solely focuses on routing messages.	Used to extend coverage without adding additional network traffic from telemetry data.	- Efficiently extends network range. - Minimizes additional network traffic.	- Does not provide its own status or telemetry data.
Sensor	Gathers and transmits sensor data, such as environmental readings. Participates in routing but prioritizes its own telemetry.	Ideal for environmental monitoring stations or any application requiring regular data reporting.	- Provides valuable data to the network. - Participates in message routing.	- May increase network traffic with frequent data transmissions.

Device Role	Description	Use Case	Pros	Cons
Tracker	Primarily used for tracking the location of assets or individuals by sending GPS coordinates.	Suitable for asset tracking, personal location monitoring, or fleet management.	- Provides real-time location data. - Assists in routing messages.	- Frequent position updates can increase network traffic.
Client Hidden	Operates in stealth mode, not appearing in node lists.	Useful for covert installations where the device should remain undetected.	- Enhances privacy and security. - Participates in routing without revealing its presence.	- May be harder to manage or troubleshoot due to hidden status.

Key Considerations:

- **Default Role:** The **Client** role is suitable for most users and scenarios.Meshtastic Node+1Meshtastic+1

- **Mobile Devices:** For portable or mobile devices, the **Client Mute** role is recommended to conserve battery and reduce unnecessary network traffic.YouTube+3Meshtastic+3Meshtastic Node+3

- **Network Infrastructure:** The **Router** and **Repeater** roles should be assigned to devices in strategic, stationary positions to effectively extend network coverage. Misuse of these roles can lead to network inefficiencies.Meshtastic

- **Specialized Functions:** Roles like **Sensor** and **Tracker** are tailored for devices with specific tasks, such as environmental monitoring or GPS tracking.Meshtastic

Selecting the appropriate device role ensures optimal performance and reliability of your Meshtastic network.

In Meshtastic, **modem presets** determine the communication parameters of devices within the mesh network, influencing both data transmission speed and range. Below is a table summarizing each preset, its characteristics, and suitable use cases:

Modem Preset	Data Rate	Spreading Factor (SF)	Band-width	Link Budget	Range	Use Case	Pros	Cons
SHORT_TURBO	21.88 kbps	SF7	500 kHz	140 dB	Short	High-speed data transfer where range is not a priority.	• Fastest data rate. - Minimizes airtime, reducing network congestion.	• Shortest range. - 500 kHz bandwidth may not be legal in all regions.
SHORT_FAST	10.94 kbps	SF7	250 kHz	143 dB	Short	Situations requiring quick data transmission over limited distances.	• High data rate. - Suitable for dense networks needing reduced airtime.	• Limited range compared to slower presets.
SHORT_SLOW	6.25 kbps	SF8	250 kHz	145.5 dB	Short	Balanced approach for moderate speed and range requirements.	• Improved range over SHORT_FAST. - Reasonable data rate.	• Still limited in range for expansive networks.
MEDIUM_FAST	3.52 kbps	SF9	250 kHz	148 dB	Medium	General-purpose use with a balance between speed and range.	• Suitable for typical applications. - Balanced performance.	• Not optimized for either extreme speed or range.
MEDIUM_SLOW	1.95 kbps	SF10	250 kHz	150.5 dB	Medium	Scenarios where extended range is needed without significant data load.	• Enhanced range. - Reliable in moderately sized networks.	• Slower data rate may not suit high-throughput needs.

Modem Preset	Data Rate	Spreading Factor (SF)	Bandwidth	Link Budget	Range	Use Case	Pros	Cons
LONG_FAST	1.17 kbps	SF11	250 kHz	153 dB	Long	Default setting; suitable for most applications requiring reliable communication over longer distances.	• Good range. - Default setting simplifies configuration.	• Slower data rate.
LONG_MODERATE	0.88 kbps	SF11	125 kHz	155.5 dB	Long	Extended range with moderate data requirements.	• Increased range due to narrower bandwidth.	• Further reduced data rate.
LONG_SLOW	0.44 kbps	SF12	125 kHz	158 dB	Long	Applications necessitating maximum range with minimal data transmission.	• Maximum range achievable.	• Very slow data rate. - Not suitable for time-sensitive data.
VERY_LONG_SLOW	0.24 kbps	SF12	62.5 kHz	161 dB	Very Long	Specialized use cases where range is critical, and data speed is negligible.	• Extends range to theoretical limits.	• Extremely slow data rate. - Not recommended for regular use due to potential network instability.

Key Considerations:

- **Spreading Factor (SF):** Higher SF increases range but decreases data rate and increases airtime.

- **Bandwidth:** Wider bandwidth allows higher data rates but may be restricted by regional regulations.

- **Link Budget:** Higher link budget indicates better signal strength and potential range.

- **Legal Compliance:** Ensure the selected preset complies with local frequency and bandwidth regulations.Meshtastic+1Meshtastic+1

Selecting the appropriate modem preset is crucial for optimizing network performance, balancing the need for speed against the requirement for range.

The **WiFi Radio** configuration option in **Meshtastic** allows the device to **enable or disable its WiFi interface**, and choose how it behaves when WiFi is enabled. This option is especially useful for **gateways**, **bridges**, or for **configuring devices without using Bluetooth**.

☑ What Does It Do?

It controls whether your Meshtastic device:

- **Connects to an existing WiFi network** (client mode),

- **Creates its own WiFi access point** (AP mode), or

- **Turns WiFi off entirely** to save power or minimize emissions.

📶 WiFi Configuration Modes (and Their Use Cases)

WiFi Mode	Description	Use Case	Pros	Cons
Disabled	WiFi radio is completely turned off.	Ideal for mobile/battery-powered nodes or stealth applications.	- Saves power - Reduces RF noise	- No OTA configuration - No bridging to internet
Client Mode	Device connects to a local WiFi network using SSID/password.	Used for internet gateway nodes, or when you want to control/view the device remotely.	- Internet connectivity - Remote updates/configuration possible	- Requires stable local WiFi - Can drain more battery

WiFi Mode	Description	Use Case	Pros	Cons
Access Point (AP) Mode	Device creates its own WiFi network you can connect to (usually "Meshtas-tic-XXXX").	Great for initial setup or when Bluetooth isn't working.	- No external WiFi needed - Easy configuration via browser	- Limited range - Only useful for nearby connections
Client + AP (Dual Mode)	Connects to WiFi and runs its own AP for nearby access.	Ideal for debugging or when you're setting up a remote node that also needs internet.	- Flexibility - Dual access options	- Higher power consumption - Can be buggy on some devices

⚙ When to Use It

- Use **Client mode** if:

    o You want to forward messages to/from **MQTT or Meshtastic cloud servers**.

    o You're using the node as an **internet bridge**.

    o You want to view metrics on the **Meshtastic web interface**.

- Use **Access Point mode** if:

    o You need to **configure the device without Bluetooth**.

    o You're in a field scenario with **no internet access**.

- Use **Disabled mode** if:

    o You're **power-constrained**, like on solar or battery.

    o You're in a **stealth** or RF-limited environment.

🧠 Pro Tip:

After enabling WiFi, you can visit the device's local IP (like 192.168.4.1 for AP mode) to **access its Web UI**. This is great for diagnostics, logs, and sending test messages.

Stationary Use Cases

Here's a breakdown of **better uses** for Mesh Devices in a **stationary context**:

☑ Best Stationary Use Cases for Screenless Meshtastic Devices

Use Case	Suggested Role	Why It's a Good Fit	Pros	Additional Setup
Router / Repeater Node	Router or Repeater	No need for a screen; just sits in a high place relaying messages	- Expands mesh coverage - Fully headless operation - Low maintenance	Configure once via USB/serial or Bluetooth; power via USB or battery
Relay Node on a Hill, Roof, or Tower	Repeater	Mounted in a weather-protected box with a good antenna for max range	- Provides long-range relays - Hidden or out-of-sight install	External antenna + weatherproof enclosure recommended
Environmental Sensor Node	Sensor	Add sensors via GPIO/UART (e.g., temperature, humidity)	- Passive data source - Useful for agricultural or home automation	Use ESP32's GPIOs; program sensor logic in firmware
Tracker Node (Still with GPS)	Tracker	GPS still works even if screen doesn't, and it can send position data	- Perfect for asset tracking - Stationary "geofence" marker	Optional: hook to external GPS antenna if needed
Headless Gateway Bridge (WiFi or Serial)	Client or Gateway	If connected to WiFi, it can forward to MQTT/HTTP servers	- Adds internet to the mesh - Ideal for hybrid mesh setups	WiFi setup via meshtastic --set wifi_ssid and wifi_password

🧠 Strategic Advantages

- **No screen = lower power draw**, perfect for long-term use.

- If you already have working **handheld devices**, use the screen-less ones to **build the backbone** of your mesh.

- They're ideal for **hard-to-reach places** — once configured, you won't need to access them often.

⚙️ Powering Options

- **USB wall adapter** for indoor/continuous power.

- **Solar + 18650 battery** for off-grid repeaters.

- Use deep-sleep modes (if not relaying) to extend battery life.

🚀 Pro Tip:

Flash them with a **custom config** and set --is_router true or --role repeater using CLI:

Alternative Use Cases

The **Meshtastic ecosystem** is incredibly flexible and can be used for **much more than messaging or mesh extension**. With the right configuration, these devices become powerful **off-grid tools**, **data collectors**, and **automation nodes**, especially since many use ESP32 boards with GPIO, I2C, UART, and analog input support.

Here's a table of **alternative uses** for Meshtastic devices, categorized by function:

🔧 Alternative Uses for Meshtastic Devices

Use Case	Role/Setup	What It Does	Hardware Needed	Why It's Useful
Environmental Monitoring	sensor	Collects & transmits data: temperature, humidity, CO2, etc.	T-Beam + DHT22, BME280, or other I2C sensors	Real-time conditions off-grid (e.g., greenhouses, field stations)
Asset or Livestock Tracking	tracker	Sends GPS location every X minutes	T-Beam (built-in GPS)	Track equipment, animals, or people in rural/off-grid areas
Weather Station Node	sensor	Measures barometric pressure, wind, rain, temperature	External weather sensors + ESP32	Set up remote, broadcast-only weather telemetry
Emergency Broadcast Beacon	client_mute + auto-message	Periodically broadcasts emergency info like SOS or location	Any ESP32 LoRa board	Passive node that repeatedly sends preconfigured emergency alerts
Tamper Detection / Perimeter Alert	sensor + GPIO interrupt	Sends a message if a door/gate is opened or a wire is cut	Magnetic sensor / reed switch / PIR sensor	Instant off-grid alerts for motion or intrusion

Use Case	Role/Setup	What It Does	Hardware Needed	Why It's Useful
Remote Logging of Events	sensor	Tracks button presses, pressure mats, etc., for time logging	Any GPIO-triggered sensor + ESP32	Useful for tracking deliveries, entry points, or workflow steps
Geofence Alert Node	tracker + logic rules	Sends a warning when an asset leaves a set GPS area	T-Beam (GPS built-in)	Protects valuable gear from being moved without permission
Vehicle Telemetry Beacon	tracker + optional sensors	Sends GPS + sensor data (voltage, fuel, temp) periodically	T-Beam + voltage divider or OBD sensor	Monitor vehicle health/status remotely
Bridge to Other Devices/Networks	client + serial/WiFi	Connects to microcontrollers (Arduino, Pi) to relay their data	UART or I2C bridge	Use LoRa to extend any IoT project's range
Offline Mesh Chat Room at Events	client	Hosts private, localized text chat at events, no internet	Any device + paired phones	Use at concerts, festivals, prepper meets, etc.
Mesh-Powered Digital Bulletin Board	client + display	Displays public broadcast messages (e.g., alerts or announcements)	E-ink or OLED screen + ESP32	Off-grid signage or announcements in public or remote areas
LoRa-Connected Doorbell or Mailbox Notifier	sensor + button	Sends message when doorbell is pressed or mailbox is opened	Momentary switch + ESP32 + LoRa	Great for cabins, farms, or remote buildings
Time-Synced Sensor Network	sensor + GPS time	Collects data across many nodes with accurate timestamps	GPS-capable nodes	Perfect for experiments or time-critical monitoring

⚡ Hardware Considerations

- Devices like **TTGO T-Beam** (with GPS & LoRa) are ideal for tracking.

- **Heltec LoRa 32 V3**, **LILYGO T-Echo**, or **RAK boards** work well for sensors.

- Some devices can be battery- or **solar-powered**, making them ideal for remote placement.

🧠 Smart Combinations

- **Tracker + Sensor:** Monitor location *and* data (e.g., GPS + temperature).

- **Repeater + Emergency Beacon:** A repeater node that also sends SOS or status.

- **Sensor + GPIO Trigger:** Use LoRa as a **low-power wireless alert system**.

RAK Device VS LoRa Meshtastic Device

Understanding the difference between a **RAK** device and a **LoRa Meshtastic device** boils down to understanding the **roles of manufacturers vs. protocol** and **how Meshtastic uses LoRa-compatible hardware**.

Let's break it down clearly:

☑ Terminology Breakdown

Term	What It Refers To	Example
LoRa	A wireless modulation technology (Long Range) designed by Semtech for low-power, long-distance data transmission	Used in sensor networks, smart cities, Meshtastic, etc.
Meshtastic	An open-source firmware that runs on LoRa-capable devices to create off-grid mesh networks	Turn your T-Beam or RAK WisBlock into a mesh communicator
RAK	A hardware manufacturer (RAKwireless) that makes LoRa-capable boards, gateways, antennas, and sensor modules	RAK4631 WisBlock, RAK7258 Gateway
LoRa Meshtastic Device	Any LoRa-capable microcontroller flashed with Meshtastic firmware	TTGO T-Beam, Heltec LoRa 32 V3, RAK4631

🆚 RAK vs. LoRa Meshtastic Device — What's the Difference?

Aspect	RAK Devices	LoRa Meshtastic Devices (Generic)
Manufacturer	Made by RAKwireless	Made by various brands (LILYGO, Heltec, etc.)
Compatibility	Many RAK devices support Meshtastic (e.g., RAK4631)	All supported boards must have LoRa + ESP32 or STM32
Form Factor	Often modular (WisBlock) or gateway-class (RAK7249)	Often integrated (e.g., T-Beam = GPS + LoRa + ESP32 in one board)

Aspect	RAK Devices	LoRa Meshtastic Devices (Generic)
Focus	Industrial IoT, sensors, gateways, modular prototyping	DIY-friendly LoRa messengers and trackers
Typical Use	Sensor networks, gateways, weather stations, agricultural monitoring	Off-grid comms, mesh networking, GPS tracking, emergency tools
Power Options	Many support solar, battery, PoE	Usually USB or battery powered (sometimes with 18650 support)
Meshtastic Use	Some RAK modules are officially supported by Meshtastic	Many devices are designed for Meshtastic use from the start

🎯 Example Devices

RAK Devices That Work With Meshtastic

- **RAK4631 WisBlock** – Tiny modular LoRa board (requires add-on base & power modules)

- **RAK7258 Gateway** – Indoor LoRaWAN gateway (not used *with* Meshtastic but for LoRaWAN)

- **RAK3272-SiP** – Compact LoRa module (requires external programming)

Popular LoRa Devices Running Meshtastic

- **TTGO T-Beam** – All-in-one Meshtastic workhorse (GPS + LoRa + 18650 battery holder)

- **Heltec LoRa 32 V3** – Compact and inexpensive, OLED screen included

- **LILYGO T-Echo** – Walkie-talkie-style Meshtastic communicator with speaker/mic

🧠 Summary

- **RAK** = A company that makes excellent **LoRa hardware**, some of which can run Meshtastic.

- **LoRa Meshtastic Device** = Any supported board flashed with **Meshtastic firmware** to become a mesh communicator.

- Not all LoRa devices are from RAK, and **not all RAK devices run Meshtastic by default** — but many can be adapted.

LoRaWAN vs Meshtastic

Many people understandably get confused about LoRaWAN and **Meshtastic** since both use **LoRa** radios and even some of the same hardware (like the TTGO T-Beam). But **they are not the same**, and knowing the **key differences** will help you avoid frustration and build the right kind of network for your needs.

☑ Summary Table: LoRaWAN vs Meshtastic

Feature	LoRaWAN	Meshtastic
Protocol Type	Standardized WAN protocol defined by the LoRa Alliance	Custom mesh protocol built by the Meshtastic community
Network Topology	Star topology: all devices connect to a central gateway	Mesh topology: devices relay messages peer-to-peer
Hardware	Requires LoRaWAN-compatible devices (sensors, nodes, gateways)	Runs on ESP32/LoRa boards (e.g. T-Beam, RAK4631)
Gateway Needed?	☑ Yes – all end nodes talk through a gateway	✖ No – nodes talk directly to each other, optionally to a bridge
Internet Required?	Usually yes (for data collection and dashboards)	Optional – works fully offline, great for off-grid use
Message Types	Short packets for telemetry (sensors, GPS)	Primarily text messages, GPS, telemetry, and sensor data
Data Direction	Uplink from devices to gateway, optional downlink	Fully bidirectional messaging
Range Optimization	Optimized for low-power, long-distance telemetry	Optimized for local communication and mesh routing
Encryption	AES-128 built-in (via network/session keys)	AES-256 (optional), end-to-end encryption between nodes
Typical Use Cases	IoT sensors, smart cities, agriculture, environmental data	Off-grid chat, emergency comms, hiking groups, prepping
Example Devices	Dragino sensors, RAK sensors, Laird gateways	TTGO T-Beam, Heltec LoRa 32 V3, RAK4631 + Meshtastic firmware

🧠 So Why the Confusion?

- **They both use the same LoRa radio technology.**

- Devices like the **T-Beam** can be flashed with either:

  - **LoRaWAN firmware** (like Arduino code or platform-specific apps), **OR**

  - **Meshtastic firmware**.

- But once flashed, a device is either **in the LoRaWAN ecosystem or** it's **running Meshtastic** — they don't interoperate.

🚫 You Cannot Use the Same Device for Both at the Same Time

You have to **choose** what role your T-Beam will play:

- **If you flash Meshtastic**, it will not talk to a LoRaWAN gateway.

- **If you flash LoRaWAN-compatible firmware**, it will not work in a Meshtastic mesh.

However, you **can reflash** the device anytime if you want to switch roles — they're very flexible that way.

☑ Use Meshtastic If You Want:

- Off-grid mesh communication (no cell towers, no gateways)

- Decentralized chat between devices

- Group coordination in the wild

- Emergency messaging

☑ Use LoRaWAN If You Want:

- Send sensor data (like temperature, air quality, water levels) to a central server

- Internet-connected dashboards and analytics

- Industrial or agricultural automation

- Integration with platforms like **The Things Network (TTN)**, **Helium**, **ChirpStack**

⌖ Final Thought

Your **TTGO T-Beam devices are extremely versatile**, but you have to **flash the right firmware** for the type of network you want to build.

Encryption on Meshtastic Networks

Meshtastic supports end-to-end encryption to ensure secure communication between devices on the mesh network.

🔐 Types of Encryption:

- AES-256: Optional but supported for high-security applications.
- AES-128: Default for LoRaWAN networks (used for comparison).

⌖ Purpose of Encryption:

- Prevent eavesdropping or unauthorized message interception.
- Ensure message integrity between nodes.
- Support private communication in emergency or sensitive use cases.

⚙ Encryption Implementation:

- Each node generates a unique keypair.
- Group and individual messaging keys can be configured.
- Firmware supports OTA updates for crypto modules.

🛠 Best Practices:

- Use strong keys (prefer AES-256 for high security).
- Keep firmware updated for latest encryption patches.
- Disable public telemetry if operating in high-risk zones.

Encryption ensures that even in a decentralized, offline environment, your communications remain confidential and protected from tampering.

Recommended tools and equipment for various grid down scenarios

Below are the recommended tools and equipment for various grid down scenarios:

◇ 1. Grid Down – Civil Unrest (Urban / Suburban)

Core Tools & Gear

Item	Description & Purpose
Meshtastic Device (TTGO T-Beam)	Core off-grid communication device with GPS and LoRa. Portable and rechargeable.
2nd Meshtastic Node (RAK4631)	Acts as a stationary repeater or backup node for extended mesh coverage.
Mini Solar Panel (10W–20W)	Foldable solar charger to power your T-Beam, phone, or power bank in the field.
High-Capacity Power Bank	20,000–30,000 mAh battery to recharge devices multiple times without sunlight.
Telescopic Mini Antenna	Compact antenna that increases transmission range when extended vertically.
Faraday Pouch / RF Shield Bag	Prevents GPS/cell signal tracking when carrying comms gear.
Minimalist First Aid Kit	Compact trauma and first-aid gear for urban emergency situations.

◇ 2. Grid Down – Natural Causes (Rural or Urban Disaster Zone)

Core Tools & Gear

Item	Description & Purpose
Meshtastic Node (TTGO T-Beam)	Primary mobile communicator with GPS for tracking, SOS, or messaging.
RAK4631 Solar Node in Waterproof Case	Passive relay or sensor node that can be placed in a weather-exposed location.
Weatherproof Solar Panel Kit (20W)	Fixed or semi-portable solar system with charge controller for sustainable energy.

Item	Description & Purpose
Flashlight + Hand Crank Radio	Dual-purpose emergency radio and flashlight with self-powering hand crank.
Outdoor Omni Antenna (5.8–8 dBi)	Weather-resistant antenna with improved mesh range, ideal for mounting on rooftops.
Dry Bag with Organizer	Waterproof storage for electronics and essentials.
Personal Locator Beacon (PLB)	Emergency beacon for transmitting distress signals with or without LoRa.
Compact Multitool & Paracord	For mounting antennas, making repairs, or securing gear in dynamic environments.

◇ 3. Grid Down – War Zone / Occupied Territory / Tactical Ops

Core Tools & Gear

Item	Description & Purpose
Screenless Meshtastic Node (RAK4631)	Stealthy, low-power node with no display. Used for hidden repeater or tracker roles.
TTGO T-Beam (Concealed)	Portable communicator, optionally worn or hidden in clothing for on-the-go use.
Covert Foldable Yagi Antenna	Directional, packable antenna for long-range, point-to-point ops.
Tactical Solar Charger (12W)	Low-profile solar panel for field charging in movement-sensitive areas.
Manual RF Antenna Switch	Allows toggling between omni and directional antennas manually in a static setup.
Burner Phone (Air-Gapped)	Used to manage nodes without SIM or tracking; isolated from primary digital ID.
Encrypted Flash Drive	Stores sensitive node configs, encryption keys, and logs securely.
Kevlar Cable or Zip Ties	For secure mounting of nodes or antennas in exposed or tactical locations.
Waterproof Silicone-Coated Pouch	Silent, impact-resistant storage for high-priority gear.

◇ Shared Must-Haves Across All Kits

Item	Description & Purpose
SMA Cables and Adapters	To ensure antenna compatibility across devices and allow extension away from enclosures.
USB-C + Micro-USB Dual Cable	Universal charging and firmware flashing cable.
Laminated Quick Reference Card	Printed guide with node aliases, channels, frequencies, emergency codes, and encryption.

Tactical Operations Center

🏕 What Is a TOC (Tactical Operations Center)?

A **Tactical Operations Center (TOC)** is a centralized command post used to coordinate communication, planning, and situational awareness during critical operations. In civilian terms, it's your **brain and nerve center**—a place where information is collected, decisions are made, and communications are routed to and from team members or resources in the field.

A TOC doesn't have to be a bunker or a tent full of screens. It can be a backpack-ready setup with a Raspberry Pi, a Meshtastic mesh node, a solar panel, and a radio—all organized to run autonomously or with minimal power and oversight.

⚠ Why Build a TOC for Grid-Down Scenarios?

1. Civil Unrest

In unstable urban environments, cell towers may be shut down or monitored.

A TOC provides **secure, private mesh communication**, allowing teams to:

- Relay real-time locations

- Issue emergency alerts

- Coordinate safe routes or supply drops

- Deploy mobile nodes to expand coverage temporarily

2. Natural Disasters (Floods, Fires, Earthquakes)

When infrastructure fails, response becomes chaotic.

A TOC allows for:

- **Rapid deployment of repeater nodes** in affected zones
- Tracking volunteers or search teams via GPS
- Sharing environmental sensor data (e.g., air quality, temperature)
- Local chat when phone/data networks are overloaded or down

3. Wartime or Occupied Territories

In high-risk zones, stealth and encryption are critical.

A TOC:

- Becomes a **hardened communication relay** to stay in touch with field agents
- Allows **directional or point-to-point signaling** to hidden teams
- Operates without internet, SIM cards, or cloud dependency
- Hosts **encryption keys, device logs, and fallback protocols** securely

💼 Portable TOC (Mobile, Pack-and-Go)

Item Name	Description	Purpose of Use
TTGO T-Beam	LoRa + GPS + ESP32 development board	Core mesh communication node; used for tracking and messaging
Raspberry Pi 4	Compact single-board computer	Runs local Meshtastic dashboard or acts as a bridge/gateway
RAK4631 Node	Low-power LoRa device with nRF52840 MCU	Deployed as a repeater or sensor node in the mesh
Omni Antenna (8 dBi)	360° LoRa antenna with extended range	General-purpose broadcasting across terrain
Yagi Antenna	Directional, high-gain antenna	For point-to-point communication over long distances
Antenna Switch	Manual switch for selecting active antenna	Allows switching between omni and directional antennas easily

Item Name	Description	Purpose of Use
Foldable Solar Panel (100W)	Portable, lightweight solar charger	Recharges batteries and devices in the field
LiFePO4 Battery (20Ah)	Lightweight, rechargeable DC power source	Stores energy from solar panels for extended operations
Power Bank (20,000 mAh)	USB-compatible battery backup	Keeps phones, T-Beams, and tablets charged
Rugged Android Tablet	Drop-resistant, water-resistant tablet	Used for Meshtastic control, map viewing, and communication
Burner Phone	SIM-free Android phone	Bluetooth interface for managing Meshtastic nodes discreetly
Faraday Pouch	RF-blocking pouch	Blocks GPS/cell signals when needed for stealth or protection
GL.iNet WiFi Router	Pocket-sized wireless router	Distributes local LAN/WiFi for dashboard access or bridging
Garmin eTrex GPS	Handheld GPS receiver	Provides accurate coordinates for tactical movement and tracking

🏠 Semi-Permanent TOC (Fixed, Hardened Node)

Item Name	Description	Purpose of Use
Raspberry Pi 4/5 + SSD	High-performance Pi with external SSD	Hosts dashboard, local server, or logs node activity
TTGO T-Beam	GPS + LoRa development board	Primary communication node, connects to Pi or operates independently
RAK7249 Gateway (optional)	Outdoor-grade LoRaWAN gateway	Used if bridging Meshtastic to LoRaWAN or The Things Network
Multiple RAK4631 Nodes	Reliable LoRa devices	Deployed as repeaters or sensor modules throughout the area
10 dBi Omni Antenna	High-gain 360° antenna	Mounted high to provide wide-area coverage for all nodes
Parabolic Antenna	Long-range, narrow-beam antenna	For directed communication to remote teams/sites
200W Solar Kit	Fixed panel solar power system	Primary off-grid power generation
100Ah LiFePO4 Battery	Deep cycle battery for extended power storage	Stores solar energy for nighttime or cloudy operation
1000W Pure Sine Inverter	Converts DC to AC power	Runs AC-based equipment from battery banks
Backup Generator	Gas/propane hybrid generator	Backup energy source in case of poor solar conditions

Item Name	Description	Purpose of Use
GL.iNet Brume/ Beryl Router	Secure Ethernet-to-WiFi router	Manages internal mesh network and wired gear
Rugged Laptop or Workstation	Hardened PC or laptop	Runs mesh mapping tools, data logs, and configuration software
Android Tablet (Wall Mounted)	Android device mounted for command visibility	Quick access to dashboard, maps, and node control
IronKey Encrypted USB Drive	Hardware-encrypted flash drive	Secure storage of firmware, logs, node configs, and encryption keys
Ham Radio Unit (Optional)	Dual-band VHF/UHF transceiver	Optional backup communications system
Weather Station (LoRa or USB)	Tracks environmental data like temp, wind, pressure	Adds data-rich capability to the TOC
RFID Locking Case	Secure gear case	Keeps sensitive electronics and encryption hardware protected

Glossary

ACK (Acknowledgment)

A signal sent by the receiver to confirm the successful receipt of a packet.

ADR (Adaptive Data Rate)

A mechanism in LoRaWAN to optimize data rate, airtime, and energy consumption.

AES (Advanced Encryption Standard)

A symmetric encryption algorithm used in LoRaWAN for secure communication.

BLE (Bluetooth Low Energy)

Low-power wireless communication used by Meshtastic devices for phone pairing.

Bandwidth

The width of the frequency band used for transmission, typically 125kHz, 250kHz, or 500kHz in LoRa.

CRC (Cyclic Redundancy Check)

An error-detecting code used to ensure the integrity of transmitted data.

Channel

A specific frequency or group of frequencies used for communication between nodes.

Client (Meshtastic Role)

A device that can send and receive messages but may not forward messages for others.

Downlink

Data sent from the network server or gateway to
a LoRaWAN end device.

Duty Cycle

A regulatory limit on the percentage of time a device can
transmit on a frequency band.

ESP32

A popular microcontroller with built-in WiFi/Bluetooth,
commonly used in Meshtastic devices.

Encryption

The process of encoding messages to protect data integrity
and confidentiality.

End Device

A LoRaWAN node that communicates with a gateway.
Also called a sensor or node.

FEC (Forward Error Correction)

A technique used to recover lost or corrupted data
without retransmission.

Firmware

The embedded software running on a LoRa/Meshtastic device.

Frequency Plan

A predefined set of frequencies allocated for LoRa communication
in a specific region.

GPS

Global Positioning System. Used by tracker nodes to determine location.

Gateway

In LoRaWAN, a device that bridges LoRa signals to an internet-connected network server.

Join Accept

The response from the LoRaWAN network server authorizing an end device to join.

Join Request

A LoRaWAN message used by end devices to request access to the network (OTAA).

LoRa

Long Range modulation technique that enables long-distance, low-power wireless communication.

LoRaWAN

A network protocol built on top of LoRa, enabling secure bi-directional communication between devices and servers.

MQTT

A lightweight messaging protocol used to transmit data between IoT devices and servers.

Mesh Network

A decentralized network where each node can relay messages for others.

Meshtastic

An open-source firmware that enables mesh communication over LoRa radios without internet.

Modem Preset

A set of LoRa configuration parameters (spreading factor, bandwidth) for performance tuning.

Node

Any device participating in a Meshtastic or LoRaWAN network.

OTAA (Over-The-Air Activation)

A LoRaWAN join procedure where the device dynamically negotiates session keys.

Payload

The actual data portion of a LoRa message.

Repeater (Meshtastic Role)

A device that forwards messages for others but doesn't send telemetry or receive messages.

Router (Meshtastic Role)

A node that forwards messages and may transmit its own telemetry.

SF (Spreading Factor)

A LoRa setting that determines the trade-off between data rate and range.

T-Beam

A LoRa+GPS+ESP32 development board commonly used with Meshtastic firmware.

TTN (The Things Network)

A public LoRaWAN infrastructure and community-operated network.

TX Power

The transmission power level of a device, affecting its range.

Telemetry

Environmental or diagnostic data (e.g., GPS, battery, temperature) sent by a node.

Uplink

Data sent from an end device to a gateway or another node.

WAN (Wide Area Network)

A type of network that spans a large area, such as LoRaWAN.

WisBlock

A modular development system from RAKwireless used to build sensor nodes or LoRa devices.

Yagi Antenna

A directional antenna that provides high gain in a specific direction for long-range LoRa communication.

www.ingramcontent.com/pod-product-compliance
Lightning Source LLC
Chambersburg PA
CBHW032018190326
41520CB00007B/532